ISBN 978-0-484-61389-7
PIBN 10633144

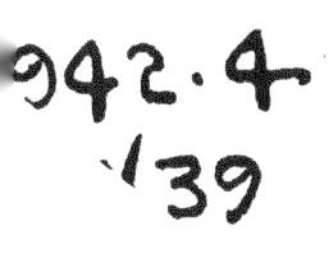

Wednesbury

As a Manufacturing & Commercial Centre and a Base for the Establishment ... of ... New Industries

OFFICIAL HANDBOOK
OF THE
WEDNESBURY
TOWN COUNCIL

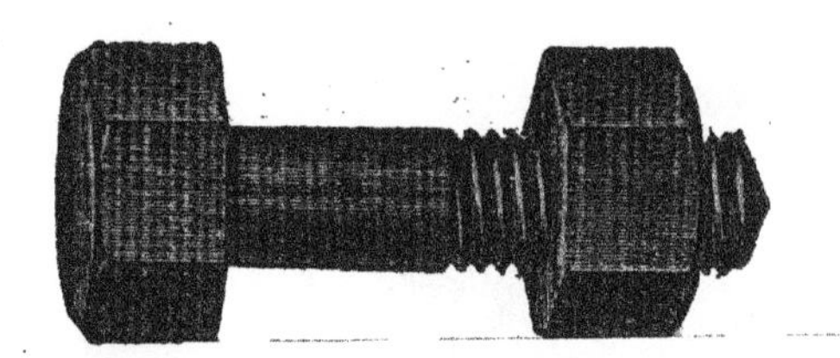

A W H B S

ETC.

WEDNESBURY

as a Manufacturing and Commercial Centre.

CONTENTS.

Also Seven Illustrations.

Official Handbook

OF THE

WEDNESBURY TOWN COUNCIL.

ED. J. BURROW & CO. LTD., PUBLISHERS,
LONDON AND CHELTENHAM.

J. W. Bernard, Photo.

The Town Hall.

Wednesbury's Commercial Advantages at a Glance.

Central Position.

Wednesbury is the very heart and centre of the South Staffordshire Black Country, being midway between Birmingham and Wolverhampton in the one direction and half-way between Walsall and Dudley in the other. It is an ancient market town, and, together with the adjoining parishes of Tipton and Darlaston, constitutes a Parliamentary Borough. The Parish of Wednesbury is in the Poor Law Union of West Bromwich, which town, till made an independent constituency in 1885, was a fourth parish included within the Parliamentary Borough of Wednesbury.

Communications.

The town is well served by a frequent service of trains on the Great Western Railway, the station on this line being about 7½ miles north of Birmingham. There is a station on the L. & N.W. Railway by which the town is reached from Walsall and the North on one side, or from Dudley Port, Worcester and the West on the other. The Midland Railway Co. have a goods station only, for which they have running powers into the town. There is good canal accommodation in the vicinity, in which there is still plenty of spare land suitable for the erection of factories, particulars of which may be had from Mr. C. W. D. Joynson, Lic.R.I.B.A., Architect, Spring Head.

Wednesbury is on the source of the Tame, which is here spanned by one of Telford's bridges on the great Holyhead Coach Road. It is also the centre of an extensive tramway system. One service, starting from Darlaston, passes through to West Bromwich, Handsworth and Birmingham; another service of cars runs into Walsall; and a third *via* Tipton to Dudley and Stourbridge.

Wealth of Raw Materials.

Wednesbury derived its earliest industrial importance from the wealth of its mineral resources, the famous ten-yard coal seam of South Staffordshire here cropping out very near the surface, a feature which led to its mines being opened out from the very earliest period of coal-getting when tools were too primitive and technical knowledge was altogether too slight to permit of output on a large scale.

A record exists of a "cole pit" in Wednesbury as early as 1315. In the reign of Elizabeth the tenants of the Manor fought out in the High Court of Chancery their rights to dig coal for their own use, and by the seventeenth century the parish was full of flourishing

" coalrys." The prosperity arose mainly through the growing demands of Birmingham's skilful smiths and artificers in metal, to whom this heavy mineral fuel had to be conveyed in panniers slung over the backs of horses; and about the middle of the following century the main reason for projecting the first canal in this Midland region was that Birmingham might be " connected with the coal works at Wednesbury " (1767).

As ironstone was always found in close proximity to these prolific and readily-gotten coal measures, it may easily be understood how South Staffordshire became the earliest and richest of England's coal and iron districts, particularly as good, serviceable fire-clay was also present, as well as lime with which to flux the metal in the furnace.

This locality has been the scene of some epoch-making experiments in the iron trade. The celebrated Dud Dudley, not having been successful in introducing his method of smelting iron with " pit coal," the next attempt was made at Wednesbury in 1686 by a German named Blewstone. The great difficulty to be overcome was the creation of a blast sufficiently strong to burn hard mineral coal in the smelting furnace: a difficulty not finally overcome till nearly a century afterwards, when the first " blast " furnace was erected by that famous ironmaster, John Wilkinson, at Bradley, near Wednesbury, in 1780. Until 1750 charcoal had been extensively used for iron smelting, to the entire depletion of all the neighbouring woodlands, when coke was substituted; but Wilkinson's ultimate success was only made possible by the employment of one of James Watt's newly-invented steam-engines to blow the blast.

Variety of Existing Industries.

Owing to its central position and the possession of such excellent communications by railway and canal the borough's industrial activity is greater than ever; and, strange to say, its products are all much heavier in freightage than they were a century back.

During the closing years of George III.'s reign, and for some considerable time afterwards, the town produced a number of small wares, like nails, wood screws, and even fine enamels, some specimens of which are still treasured by collectors and in museums. But at the present time the productions consist mainly of railway wheels and axles, and other heavy railway ironwork; boiler plates, bridges and girders and general engineering work; machinery, shafting and foundry-work in brass or malleable iron; bolts, nuts, plugs, taps, cocks, ship fittings and wrought-iron work of every description; coach springs, coach axles, and general van, cart and coach ironwork; spades, shovels and heavy edge tools; pots, pans and hollow-ware; a great variety of smaller articles of hardware; various chemicals; and last, though

not least, every possible description of wrought-iron tube for conveying gas, water or steam, and the fittings and all other appliances used in wrought-iron tubing work.

Of these industries, the two leading ones are the production of railway plant and the manufacture of tubes, both of which may be said to be indigenous and evolutionary. When, about a century ago, gas-lighting first came into vogue, this town had a reputation for the manufacture of gun barrels. The conveyance of the new artificial illuminant was at first a difficulty; it was carried in copper tubing, in brass tubing, and even when iron gun barrels screwed end to end were used, the cost was excessive. To obviate this, an ingenious Wednesbury workman, named Cornelius Whitehouse, after many discouragements, at last succeeded in making a cheap wrought iron gas-pipe, which he accomplished by taking a long strip of iron, heating it in a special elongated furnace, and manipulating it entirely by machinery of his own devising. The cheapness of these machine-welded tubes, as compared with the cost of the elaborate process by which short gun barrels were produced, at once revolutionised the trade in iron pipes.

The Labour Market.

Wednesbury is the centre of a thickly populated area, and the supplies of skilled and unskilled labour for the light and heavy trades are ample. At the present time, of course, there is a grave shortage of male labour.

The Advantage of a Knowledge of Heraldry.

As the vital interests of this industrial borough are thus so exclusively wrapped up in coal and iron, it is not surprising to find " canting " allusions to the fact in the Borough Arms. Nominally the heraldic device of Wednesbury is an adaptation of the coat-of-arms of the Heronville family, who were Lords of this Manor in the twelfth century. But on the fesse in the middle of the shield there is " the symbol of the planet Mars between two sable lozenges "—the latter being easily interpreted as two " black diamonds," otherwise coal; while the symbol itself was also used by the alchemists of old to denote iron, " the metal of Mars," which classic deity, again, was identical with Woden, the Saxon God of War, from whom the town derives its name. The crest, " a tower in flames," is as near to a blast furnace as the fantastic science of heraldry can approach; and in the motto, *Arte, Marte, Vigore,* translated into " By skill, by iron, by energy," the term *Marte* has a second meaning " By arms," in allusion to the old staple industry of Wednesbury, the now obsolete gun trade. So that there is some significance in a shield of arms if one only knows how to interpret it.

Supplies of Gas, Water and Electric Light.

Gas for power and lighting is supplied by the Birmingham Corporation, and prices for large and small consumers can be obtained from the Gas Offices, in Bridge Street. The present War prices are no criterion of what obtain in normal times.

The growth of the electricity undertaking has been phenomenal in recent years, this being partly due to the fact that practically every factory and works has scrapped old methods and insisted on the installation of electric drive for power. Almost every class of drive for machinery is supplied by the Electricity Department, and the Engineer and Manager is only too pleased to advise potential manufacturers or private consumers with regard to plant for lighting, heating or power. (See page 48.)

There are over sixteen miles of mains laid, connected with which are over 30,000 lamps and motors of over 3,000 h.p. in the aggregate, but ranging in size from the humble 1-8 h.p. to the mighty 200 h.p.

The undertaking is in a very satisfactory position, there being plenty of power available, and still further plant being put down. Special terms are offered to long-hour users and to those who wish for complete electric-light equipment in their private residences.

Technical Education.

The Science School is a small building wedged in between the Post Office and the Education Offices; but it is well-equipped with laboratories and furnaces, and if judged by the excellent results it achieves in the teaching of metallurgy, metal working, the study of iron and steel, and all the kindred subjects so vital to the commercial and manufacturing interests of the locality, it takes very high rank among the Science Schools of the whole kingdom. In this connection see, also, the announcements relating to the Birmingham Corporation Industrial Research Laboratories on pages 00.

The Municipal Technical School was founded in 1892, and is housed partly in a building specially erected for the purpose and partly in certain rooms within the Art Gallery buildings. The School specialises in many subjects, and is under the direction of a Higher Education Committee.

The Metallurgical and Engineering Institute. These buildings were completed and opened directly after the outbreak of the War, viz., in September, 1914. The buildings are extensive and splendidly equipped, and have done excellent work in training both sexes for munition work. The institution is maintained by the Staffordshire County Council.

BIRMINGHAM CORPORATION INDUSTRIAL RESEARCH LABORATORIES.

Gas Department, Council House, and at Windsor Street Gas Works, Birmingham.

ENGINEER IN CHARGE .. MR. C. M. WALTER, B.SC., A.M.I.E.E., M.INST.MET.
ASSISTANT ENGINEER .. MR. V. E. GREEN, A.M.INST., C.E., A.M.I.Mech.E.

The Staff engaged are specialists in their particular departments.

THE GAS SUPPLY.

The Gas Supply authority for Wednesbury and Darlaston is the City of Birmingham Gas Department, whose Branch Office is situated in Bridge Street, Wednesbury.

Gas consumers in the district have the opportunity of consulting any of the Birmingham Gas Department Advisory Experts in connection with the use of Town's gas for all purposes.

RESEARCH LABORATORIES.

Research Laboratories have recently been established with a view to collaborating with manufacturers in carrying out work of a scientific and investigational character.

Problems must at times present themselves to manufacturers in Wednesbury when certain scientific research would be of considerable assistance to them, and it is hoped that in any difficulty which may arise they will not hesitate to consult the experts who have been specially engaged to conduct investigations at the above Laboratories.

RESEARCH WORK.

Research work is undertaken without any regard to its effect on the sales of gas, and is of the greatest possible variety. The work carried out not only deals with the use of gas, but includes investigational work as to the composition of metals, the heat treatment of steels, brasses, bronzes, aluminium, and the many special alloys which enter into local manufactures, and physical and mechanical testing, for which plant has been erected, and which would bear comparison with any in the country.

Research work relating to electrical appliances, glass-ware, and problems which arise from time to time in the manufacture of articles of non-ferrous and ferrous metals, can be dealt with in a very comprehensive manner.

A considerable amount of special investigational work has been carried out with the aid of the photo-micrographical apparatus which has been installed in the laboratories for examinations of steels, brasses, bronzes, and other alloys.

VALUE TO MANUFACTURERS.

The success of the undertaking in assisting manufacturers is beyond question. Discoveries have been made in these laboratories **which have saved manufacturers in Birmingham many thousands of pounds.**

Manufacturers should have no hesitation in bringing to the laboratory work of a confidential nature, and they can rely upon the utmost care being maintained throughout.

THE EXPERIMENTAL FOUNDRY.

At the Windsor Street Works of the Gas Department are situated the experimental foundry and hardening shops, having a floor space of about 1,700 square feet, which are used for the preparation, heat treatment and melting of metals and alloys, and which have been equipped with special plant to enable manufacturers to carry out on a full manufacturing scale any tests they may require in connection with annealing, carburizing, re-heating and quenching, normalizing, hardening and tempering of fine tools, gauges, cutters, beds, dies, the heat treatment of high-speed and alloy steels, annealing of cast iron, and the manufacture of malleable castings, the melting of brass, bronze, gun-metal, white metals, aluminium and aluminium alloys.

For the hardening and tempering of very large steel dies, beds, tools, cutters and machine parts, special plant has been installed.

Any further particulars will be readily given on application to the Engineer in charge.

[See page 3 of Cover.]

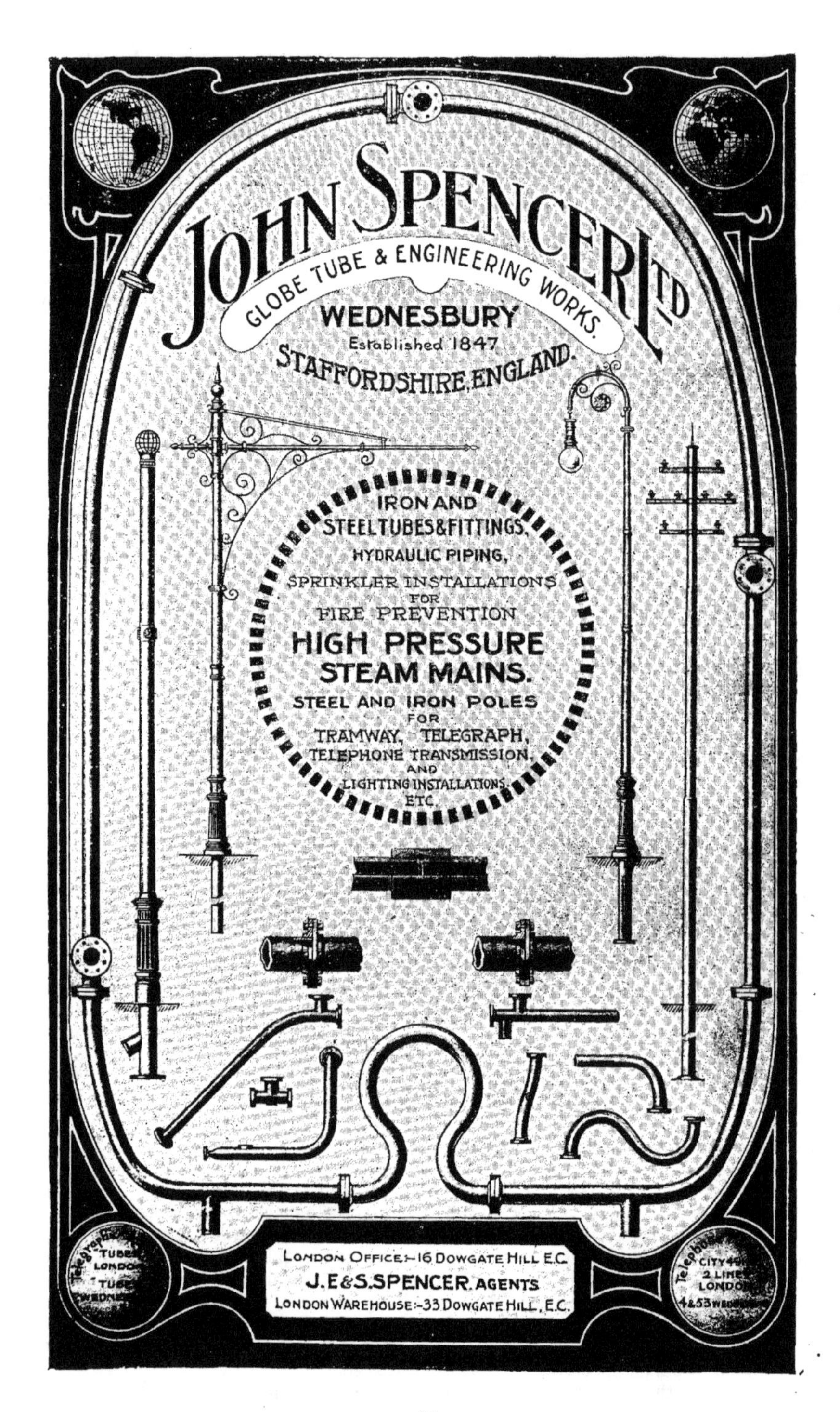
JOHN SPENCER LTD
GLOBE TUBE & ENGINEERING WORKS.
WEDNESBURY
Established 1847
STAFFORDSHIRE, ENGLAND.
IRON AND
STEEL TUBES & FITTINGS,
HYDRAULIC PIPING,
SPRINKLER INSTALLATIONS
FOR
FIRE PREVENTION
HIGH PRESSURE
STEAM MAINS.
STEEL AND IRON POLES
FOR
TRAMWAY, TELEGRAPH,
TELEPHONE TRANSMISSION,
AND
LIGHTING INSTALLATIONS
ETC.
LONDON OFFICE:–16 DOWGATE HILL E.C.
J. E & S. SPENCER. AGENTS
LONDON WAREHOUSE:–33 DOWGATE HILL, E.C.

The Tube Industry.

When one comes to look into the matter, tubes now have such varied and such general uses that it would be next to impossible for the urban resident to get on without them. He could not enjoy that luxurious bath first thing in the morning; could not have his breakfast cooked by gas or electricity; could not go off to his business by train or cycle; could not have his office heated by hot water; could not switch on his electric light when he wanted to put in a little overtime; could not use either the telegraph or telephone; could not pay business calls by tram-car; and could not use a really up-to-date bedstead when he retired to his well-earned repose.

If he is engaged in mercantile affairs on a large scale, his debt to the tube-makers is so much the more increased. The locomotives that transport his raw materials and finished manufactures; the ships that bear his goods to many lands; the concerns that supply his factories with light and power; the commercial vehicles that serve him on the high roads; and the boats and barges that ply with his precious freights along the canals—all would be crippled and helpless if denied aid from the makers of tubes.

From the first the industry has never been without golden opportunities and new realms to conquer. The universal adoption of steam for driving engines and the universal adoption of gas for household and street lighting, were in themselves enough to set any trade on its legs. But after these came greater chances offered by the triumphs of electricity and the rapidly-expanding motor business, to say nothing of the wide-spread insistence on improved sanitary conditions in all self-respecting towns. One can easily conceive, then, that with wonderful development in all these spheres the great tube-making firms of Wednesbury found themselves on the full flood of fortune's tide—works were extended only to be pulled down in a short time to rise roomier and better-equipped than ever before. New methods and new tools were adopted, and, with practically no limit in the demand (especially from foreign and Colonial buyers), foremen and employees were taught how to multiply their output to a degree that would have startled the trade pioneers of the eighteenth and nineteenth centuries.

While there is no lack of initiative on the part of individual firms, there are certain definite principles to which each adheres, and had we the space to spare, a description of the numerous processes involved would make interesting reading.

A glance through the following section and the corresponding advertisement pages will show the extraordinary variety of the tubes manufactured, nor should it be forgotten that a very large business is done in flanges, plugs, cocks, valves and other fittings, also in sanitary flush pipes of galvanised steel.

Tube galvanising by special processes is a branch of the industry that employs many hands.

REPRESENTATIVE TUBE MANUFACTURERS.

James Russell & Sons Ltd., Crown Tube Works, Inventors, Patentees and First Manufacturers of Gas Tubing.

Messrs. James Russell & Sons Ltd. very rightly claim the distinction of being the pioneers in the Welded Tube trade, for the originator of this firm, James Russell, was the first to invent and manufacture gas tubes. In the early days the method of manufacture must have been on crude lines, for gas tubes were then made from gun-lock barrels ; and it is worthy of mention that at the present time a number of trade dealers of the old school continue to use the designation "gun barrels" when ordering gas tubes.

This firm has a reputation for supplying high-class quality gas, water, steam and hydraulic tubes and fittings, black and galvanised, also steam mains, well boring and oil well tubes, etc. ; and their well-known "Crown Brand" has been supplied to buyers in every corner of the globe.

Messrs. James Russell & Sons Ltd. have also extensively manufactured light lapwelded steel tubes with inserted joints, for lead and yarn, and similar tubes with "Kimberley" collar joints. The latter joint was designed and first used for the Kimberley water mains in South Africa, and James Russell & Sons Ltd. received the major part of this order, which was regarded at the time as the largest contract ever given out in the tube trade. This Company supplied large water mains of the same description for Port Elizabeth, South Africa ; Vancouver, British Columbia ; and elsewhere.

The firm are large makers of tapered and sectional poles for tramways, telegraphs, and other transmission purposes, as may be gathered from the fact that the poles and accessories for many installations of tramway equipment in this country were made by them, including those at Leeds, Hull, Blackburn, Newcastle-on-Tyne, Salford, Carlisle and Rochdale ; and abroad at Madrid, Barcelona, Camps Bay (South Africa) and Hong Kong.

The Company's chief manufactory is at the Crown Tube Works, Wednesbury, where the fully-equipped premises include a large Brass and Screwing Tackle Works, while close at hand they have their King Street Coil Works. Their important Galvanizing Works are at Darlaston, and, in addition, they have Depots at London, Leeds, Manchester and Birmingham.

[See Advt. p. 34.]

Thomas Pritchard Ltd., Manufacturers of Wrought-Iron Tubes and Fittings of every description, South Staffordshire Tube Works, Mestycroft, Wednesbury.

This firm was established in 1857, and the founder, the late Thomas Pritchard, was one of the pioneers of the Tube and Fittings industry, having brought out several patents: notably that for making Welded Iron Bedstead Tubes. He retired from business in 1886 and the Works subsequently passed into the sole direction of his son, Mr. Albert E. Pritchard.

Through his untiring and watchful activity and resolute purpose to maintain a high standard of efficiency, the works have been extended from time to time, and now cover a considerable area at Mestycroft.

The business has grown considerably, and lapwelded tubes for boilers, and all kinds of tubes and fittings for Gas, Steam, Water and Hydraulic purposes are produced in large quantities.

Apart from the exigencies of business, Mr. Pritchard has devoted no inconsiderable portion of his time (over 25 years) to matters of Local Government, and holds many important positions, he being a J.P. for the Borough, a County Magistrate, an Income Tax Commissioner, and the present Mayor of Wednesbury.

[See Advt., p. 4 of Cover.]

REPRESENTATIVE MANUFACTURERS.

John Spencer Ltd., Globe Tube and Engineering Works.

The Globe Tube Works were founded by Mr. Cornelius Whitehouse, who was among the first inventors of one of the present processes of making Buttwelded Tubes, his patent having been taken out in February, 1825.

Mr. John Spencer, who was originally at the Vulcan Tube Works, West Bromwich, acquired the Works of Messrs. Whitehouse & Co., and since then great developments have taken place.

John Spencer Limited are large makers of **Lapwelded and Buttwelded Iron and Steel Tubes** of all descriptions, and they are specialists in all kinds of **High-Pressure Steam and other pipes.**

They are manufacturers of Poles for Tramways, Telegraph and Telephone purposes. Their poles have been employed exclusively by many of the great Public and Private Tramway Undertakings in all parts of the world. They also make Taper Telegraph Poles of all sorts.

John Spencer Ltd. have numerous Experts, who can advise on all matters connected with Pipe Installations, and who can undertake the design, supply and erection complete where desired.

They also supply complete Installations of **Automatic Sprinklers** for Fire Prevention, and have equipped a large number of Factories, Munition and Explosive Works with special apparatus. It is worthy of note that their Sprinklers secure the highest rebates from Insurance Companies.

[See Advt. p. 10.]

James McDougall Ltd., Hope Patent Tube Works, Mestycroft.

One of the leading industries of Wednesbury to-day is the manufacture of Tubes and fittings, and eminent among establishments of its kind is that conducted under the above-mentioned style. The business was founded by Mr. James McDougall in 1869, and since his death, which occurred in 1888, the concern has been most ably conducted by his representatives—Messrs. James McDougall (Managing Director) and Daniel Tonks (Director). This is one of the largest establishments of the kind in the district, and during the whole of its existence it has been in the full enjoyment of a wide patronage and high reputation. The business is ably conducted in magnificently-appointed premises. The Works combine ample space, with a very convenient arrangement of different departments. There are facilities for the manufacture of bedstead, blind and fencing tubes, with ferrules for same; also oval, flat and section tubes of all descriptions; special light tubing; extra strong tubes, handrail and blind tubes, etc.; gas, steam and water tubes and fittings; cold-drawn weldless steel tubes for cycles (which are specially prepared of a quality having great strength and toughness as required for the trade), seamless steel tubes for hydraulic and engineering purposes, etc., etc. The appliances in use are of the most approved modern design, and the many orders are executed with fidelity and despatch. The business is most comprehensive in its detail, and the productions of the firm are widely known, being welcomed in every market to which they are introduced. By superiority of material and excellence of workmanship the high character of the business has been maintained by the late founder's representatives.

In the year 1917 the firm acquired the business of Messrs. Isaac Griffiths & Sons, Imperial Tube Works, Wednesbury. This is a very old-established concern, with well-appointed premises and Offices for the carrying on of their well-known manufactures of Gas, Steam, Water Tubes and Fittings, also Imperial Steel Conduit Tubes and accessories. Further improvements have since been made to meet the increased demand for this class of tubing, and every effort is made to execute orders with promptitude and despatch.

[See Advt. p. [illegible]]

Edward Smith Ltd., Brunswick Tube Works,

supply the Shipbuilding, Gas, Electrical, Automobile and General Engineering trades, also Railway Companies and Public Contractors, with high-quality tubes and fittings in iron and steel. Their manufactures comprise butt-welded and lap-welded iron and steel tubes for gas, water, steam, compressed air, various hydraulic purposes, ships' pillar tubes, boiler and stay tubes, well boring and lining tubes, roller and loom tubes, point rods, etc.

The firm also supplies general castings to customers' patterns and drawing.

Established in 1850, the business has won a first-class reputation for reliability and quality of products. [See Advt. p. 37.]

Axles and Springs.

Here, again, things commonplace yet vitally important in their own sphere call for consideration.

In what have been termed " the good old days " a big trade developed at Wednesbury in the manufacture of axles and other fittings for the coach-builders and wheelwrights.

The present flourishing trade in fittings for railway engines and carriages has afforded more than ample compensation for the decline of the older industry.

At first, axles used on railway carriages were made exactly like the coach axles, except that they were of larger dimensions. As they sometimes broke down under the severe strain imposed upon them by the greater speed and weight of the modern rolling stock, the mechanicians directed their attention at an early stage towards the problem of strengthening the fibre of the axles.

To the Rev. James Hardy, a Baptist minister at Wednesbury, is due the credit of suggesting the " faggotted axle " of forged iron—the idea is said to have occurred to him while noticing the divisions of an orange when it is cut transversely—and by 1835 was founded the world-famous Patent Shaft and Axle-tree Co., which some few years ago was absorbed into that great manufacturing combine, the Metropolitan Amalgamated Railway Carriage and Wagon Co. Ltd.

It was at the old Park Works of the Patent Shaft Co. that the earliest experiments in Bessemer steel-making were carried out. In 1882 the Gilchrist basic process of steel-making was first tried, and Mr. Andrew Carnegie's firm first saw the experiment in the Basic Open Hearth process here.

The advances made in methods of producing and using steel naturally affected the axle-makers, who soon found it to their advantage to exchange faggotted iron for the lighter and more reliable material.

For the greater part of the Victorian era the manufacturers of " coach-ironwork " and fittings were " in clover," for only horse-drawn vehicles were on the roads. The farmer had his dog-cart and great lumbering carts and wagons; the town dandy hired his hansom; the governess had her trap or "tub." The 'bus (with its indispensably witty driver) became a popular means of conveyance in London and the large provincial towns.

The prophecy of "Mother Shipton" with regard to horseless carriages was regarded as a rich joke, for the trade seemed secure from change.

Nevertheless, changes came, which for a while completely upset calculations; but the local manufacturers were always ready to adapt themselves to new circumstances and new conditions. Hence while a good deal of high-class work is done in making axles and springs for all kinds of road vehicles, the principal firms specialise in springs, wheels and tyres for the motor industry, and do an immense amount of business in motor body fittings.

The various works devoted to these purposes will be found equipped with the very latest plant and machine tools, and the casual visitor would be astonished to discover what huge stocks are carried in the spacious and well-organised warehouses.

REPRESENTATIVE MANUFACTURERS.

A. H. Taylor (Springs) Ltd., Spring and Axle Manufacturers, Bridge Street.

This business was established in 1904 by A. H. Taylor and incorporated into a private Limited Company in 1910.

The Founder was made Managing Director, and still retains that position. He has worked through every branch of the trade for experience, and all orders get his personal supervision.

The firm's specialities are Springs for Motor and Transport Vehicles (every class of spring being manufactured and warranted made of best material).

Connections: United Kingdom and markets in most parts of the world; Contractors to H.M. Government. [See Advt. p. 38.]

Edwin Richards & Sons Ltd., Axle Manufacturers, etc., Portway Works.

Established over a century ago by Henry Richards, this business has moved with the times and has never failed to seize opportunities for expansion as they came. The Works were re-built by Mr. Edwin Richards in 1868, and for many years after that date they were engaged on the production of axle springs and coach ironwork.

With the advent of motor-cars much of the old business connection disappeared, and re-organisation of plant and staff led to an astonishing development of trade along the new lines.

Since the outbreak of War the firm has been almost exclusively engaged on large War Office Contracts for the many pattern of axles required by the transport service. [See Advt. p. 39.]

Adams & Richards, Iron and Steel Stockholders, Manufacturers of Wheels, Tyres, Axles, Springs, etc., Bridge Street.

In their extensive warehouses Messrs. Adams & Richards carry large stocks of iron and steel of all descriptions: rounds, squares, flats, half-rounds, one-round edge, fullered angles, etc., socket and fitting iron a speciality.

They are also prepared to carry out haulage contracts on reasonable terms.

[See Advt. p. 41.]

R. Disturnal & Co., Axle and Spring Manufacturers, Brass and Iron Founders, etc., Bridge Works.

" By hammer and hand all arts do stand," says the old proverb, and the suggestive trade-mark of this firm (see page 40) is a familiar one in the Wednesbury district, recalling nearly a century of strenuous labour and achievement. " Hammer and hand " may have had much to do when the business was established so many years ago, but nowadays labour-saving machinery is in evidence at the Bridge Works, and the output of axles and springs for all kinds of road vehicles has kept pace with the big demand.

Motor manufacturers should make a note of the fact that Messrs. Disturnal & Co. are makers of first-class head and hood fittings for motor bodies; also door-hinges, locks, seats, handles, press and turn buttons, step brackets, etc.

Enquiries invited. [See Advt. p. 40.]

Engineering and Kindred Trades.

Nothing could be more striking than the contrast afforded by the huge bridges and the framework for equally huge buildings which are produced by the constructional engineers, and the various bolts, nuts, washers, set pins, studs and screws which help to keep these fabrics together. The effect of this contrast may be fully obtained at Wednesbury.

In a former Guide to the town reference was made to a local firm's achievement in the building of the Tugela Bridge during the great Boer War—the whole of the material being produced from the crude pig-iron at their own works, and the structure being erected well within the specified time.

Many striking successes have been obtained by Wednesbury's constructional engineers since then, some of the finest bridges in the world having been produced in sections here.

One of the largest engineering firms specialises in the manufacture of shafting and general power transmission machinery; also in complete plants for making tubes, bright bars, bolts, nuts, rivets and spikes, and drop-forging plant; another is celebrated for its lifting jacks, tube cutters, parallel vices, cramps and other tools for joiners and engineers; while yet a third is renowned for its shafting collars, hacksaw machines, and hacksaw blades.

These are mentioned in order to show that the prosperity of one local industry is bound up in that of another ; and that what one may call the staple trades have no occasion to go outside the town either for complete primary installations or for the renewal of plant.

Equally important in keeping the industrial machine at a high pitch of efficiency are the efforts of those manufacturers who produce iron and steel bars and strip ; and those electrical engineers and fitters who devote special attention to mechanical repairs.

With the extraordinary development of the motoring and general engineering trades, and quite recently of the aircraft industry, the local manufacturers of small accessories such as bolts and nuts have quadrupled their output, which, in the aggregate, amounts to many thousands of tons per month.

In the washer and bolt-making works will be found equipment of the most up-to-date character—the manufactures being disposed of to aircraft makers, wagon and coach builders, motor firms, railway companies, shipbuilders and colliery proprietors in all parts of the world.

Enormous demands are made on those firms that produce bright drawn steel rounds, squares, hexagons, etc., for machinery; also castings in malleable iron and steel. Light grey iron castings of superior quality are made for the engineering and electrical trades; and the iron and brass founders are kept busy (the latter more especially on gas, steam and sanitary fittings).

Many of the engineering firms are engaged on Government contracts, and to that extent, of course, cannot oblige their usual clients as they desire to do. Orders are executed, however, as promptly as the abnormal conditions will allow.

REPRESENTATIVE FIRMS.

Samuel Platt Ltd., Engineers and Machine Tool Manufacturers, King's Hill Foundry.

This is one of the most representative engineering firms at Wednesbury, whether regarded from the standpoint of age, the variety of manufacture or the number of workpeople employed.

The full story of this great business embraces the whole of four reigns and portions of two others, and the outlook for the future is distinctly promising.

In splendidly-equipped premises Messrs. Platt produce first-class shafting and shaft fittings (with patent self-oiling bearings) ; pulleys and mill-gearing ; tube-manufacturing machinery ; bolt and nut-making machinery ; reeling and straightening machines for bars and tubes (both hot and cold) ; patent drop stamps and stripping presses ; patent power hammers ; and lathe chucks—a ready sale being obtained for all these in home and foreign markets.

At the present time the firm, under Government control, is engaged solely on War Office and Admiralty Contracts,

The Works are extensive and well-organised, and the number of employees is about 2,000. [See Advt. p. 42.]

W. Wesson & Company Limited, Victoria Iron and Steel Works.

The Victoria Iron and Steel Works of Messrs. W. Wesson & Co. are entirely engaged on producing Iron and Steel for Munitions of War. The manufactures include Iron and Steel Bars for the production of Chains, Cables, Nuts and Bolts, etc; Iron and Steel Strip for Gas, Bedstead, Hydraulic and Conduit Tubes; special Puddled and Ball Furnaced Iron and Steel Billets for Stamping and Forging. It should be noted also that this firm specialises in the production of Bright, Cold-Rolled Steel Strip and Bright Drawn Steel Bars.

As Galvanizers, Messrs. Wesson have an excellent reputation in the trade: Tubes, Fittings, Nuts and Bolts, and General Oddwork being successfully dealt with by this process.

The firm has been on the Admiralty list for a considerable period, and has many Government Contracts running at the present time.

[See Advt. p. 46.]

The Steel Nut and Joseph Hampton Limited.

The illustrations on page 43 will give some idea of the activities of this Company, which has a world-wide connection—Agencies being established in the Colonies and on the Continent.

The Nut and Bolt Department, which supplies general and electrical engineers, aircraft manufacturers, motor and cycle makers and others with best-quality bright bolts, nuts, washers, studs, turn-buckles, and numerous other articles turned from the steel bar; the Steel Department, which produces drawn steel in a variety of sections for machining; the Tool Department, where carpenters', joiners' and engineers' tools are manufactured; the Foundry, which produces tramway and engineering castings—each is equipped with modern machinery enabling it to cope with contracts from the War Office, the Admiralty and the Air Ministry, which at present are absorbing most of the Company's energies.

[See Advt. p. 43.]

Hampton & Beebee Ltd.

is a name well known in the automobile, aeronautical and engineering trades, for the firm has given particular attention to the production of superior washers —parts small in themselves, but of vital importance in "keeping things together." These washers are made in a variety of metals and sizes, and are esteemed for their reliability.

Another department of the business is devoted to the manufacture of shafting collars, hacksaw machines, and hacksaw blades; and an extensive trade is done in bright steel bolts, nuts and studs, square hole grub screws, and bright steel strip.

Enquiries esteemed from the trades mentioned.

[See Advt. p. 44.]

Henry Peace Ltd., Electrical and Mechanical Engineers and Brass Founders.

The pretty general adoption of electricity as a motive power in the Wednesbury workshops has given scope for the display of enterprise, resource and adaptability on the part of the electrical and mechanical engineer, and the above firm has not been slow to take advantage of the opportunities thrust in its way.

Good business is done, also, in the production of phosphor bronze and gun-metal castings **machined to customers' requirements.**

Enquiries invited from the aircraft and motor trades.

[See Advt. p. 50.]

H. J. Barlow & Co., Manufacturers of bright steel nuts, bolts, set pins, engineers' washers, etc.,

whose works are to be found in Mounts Road, with Offices in Bridge Street, although a comparatively young firm, are very quickly making a reputation for themselves amongst the greatest engineering concerns, including the motor and aircraft factories, of this country.

They believe in giving a GOOD ARTICLE EVERY TIME, and their Works are well-organised and are equipped with the very latest type of machinery.

[See Advt. p. 45.]

Frost & Sons (Moxley) Ltd., General Galvanizers to the Trade, Falcon Works, Moxley, and Rough Hills Works, Wolverhampton.

Galvanizing is the name given to the process of coating iron with zinc to protect it from rust—zinc being electro-positive to iron and corroding before the latter is affected. The iron has to be cleaned with great care by acid treatment and scouring with sand, and is then dipped into a bath of molten zinc covered with ammonium chloride to act as a flux. For special purposes the zinc is also "deposited" by means of ingenious electric apparati.

Messrs. Frost & Sons have been in the business nearly thirty years and make a speciality of galvanizing tubes by a patent process. They have a very extensive connection in the Wednesbury and Wolverhampton districts.

Prompt attention is given to all enquiries.

[See Advt. p. 47.]

C. Walsh Graham Ltd., Timber and Slate Merchants, Joinery Manufacturers and Roof Erectors, Wednesbury, Wolverhampton and Smethwick.

This Firm, established in the same name in 1871, is of very considerable importance in Wednesbury, and has large establishments at Wolverhampton and Smethwick (Birmingham). The Directors are P. Lee, A. E. Phillips, Lieut. H. A. Jowett (Military Cross, now in Salonica), D. Harford, and G. F. Butler.

They operate an absolutely modern plant at each of their three establishments, all electrically driven ; and besides holding very heavy stocks as merchants, both in the Midlands and at the ports, they are in a position to manufacture and erect any kind of high-class finished woodwork. They are also the inventors and manufacturers of the patent "Primus" puttyless glazing, in which an ingenious lead cap clothes the bar, so that nothing but lead and glass is exposed to the weather.

Improved type lattice girder roofing is an important feature of the business, and many fine photographs of large roofs erected by them are to be seen at their Offices. Inspection is invited.

[See Advt., p. 52.]

A FIRM FAMOUS FOR HIGH-CLASS CHEMICALS.

Chance & Hunt Ltd., Manufacturers of Chemicals, Oldbury (near Birmingham), Wednesbury and Stafford.

In the various trade processes illustrated at Wednesbury, and more especially in the preparation of metals for diverse uses in the iron and steel industry and in the galvanising works, chemicals play a most important part. A list, suggestive rather than exhaustive, of the chemicals manufactured by the above well-known firm will be found on page 49.

Intending purchasers of any chemical product would do well to address their enquiries to either the firm's Head Office at Oldbury or to the London Office at 61 and 62, Gracechurch Street, Cornhill, E.C. 3.

THE MODERN BOROUGH.

The town is purely industrial, the only residential quarter being Wood Green, a suburb on the Walsall Road, with an outlook towards the breezy heights of Barr, beyond which lies the wide open expanse of Sutton Coldfield Chase.

Wednesbury was incorporated in 1886, the first Mayor being Alderman Richard Williams, J.P., for many years previously Chairman of the Local Board of Health, and who is still active in the public service. The Local Board was called into existence in 1851, owing to the insanitary state of the town, which had been dreadfully ravaged by the cholera visitations of 1832 and 1848. The ancient Manorial Court Leet became defunct upwards of half-a-dozen years ago.

The Parliamentary Borough (including Tipton and Darlaston) has 12,733 electors on the Register, and is represented by Sir John Norton Griffiths (Conservative). County voters are included in the Handsworth Division. The town sends two representatives to the Staffordshire County Council, one representing the Central Division (comprising Market Ward and Town Hall Ward) and the other the Suburban Division (comprising Kings Hill Ward and Wood Green Ward). Representatives for many years past and still representing the borough, A. E. Pritchard, Esq., J.P., and C. W. D. Joynson, Esq., J.P.

The town is policed by the Staffordshire Constabulary; the County Magistrates hold a Petty Sessions, alternately with the South Staffordshire Stipendiary, every Tuesday; and the Borough Magistrates sit every Friday.

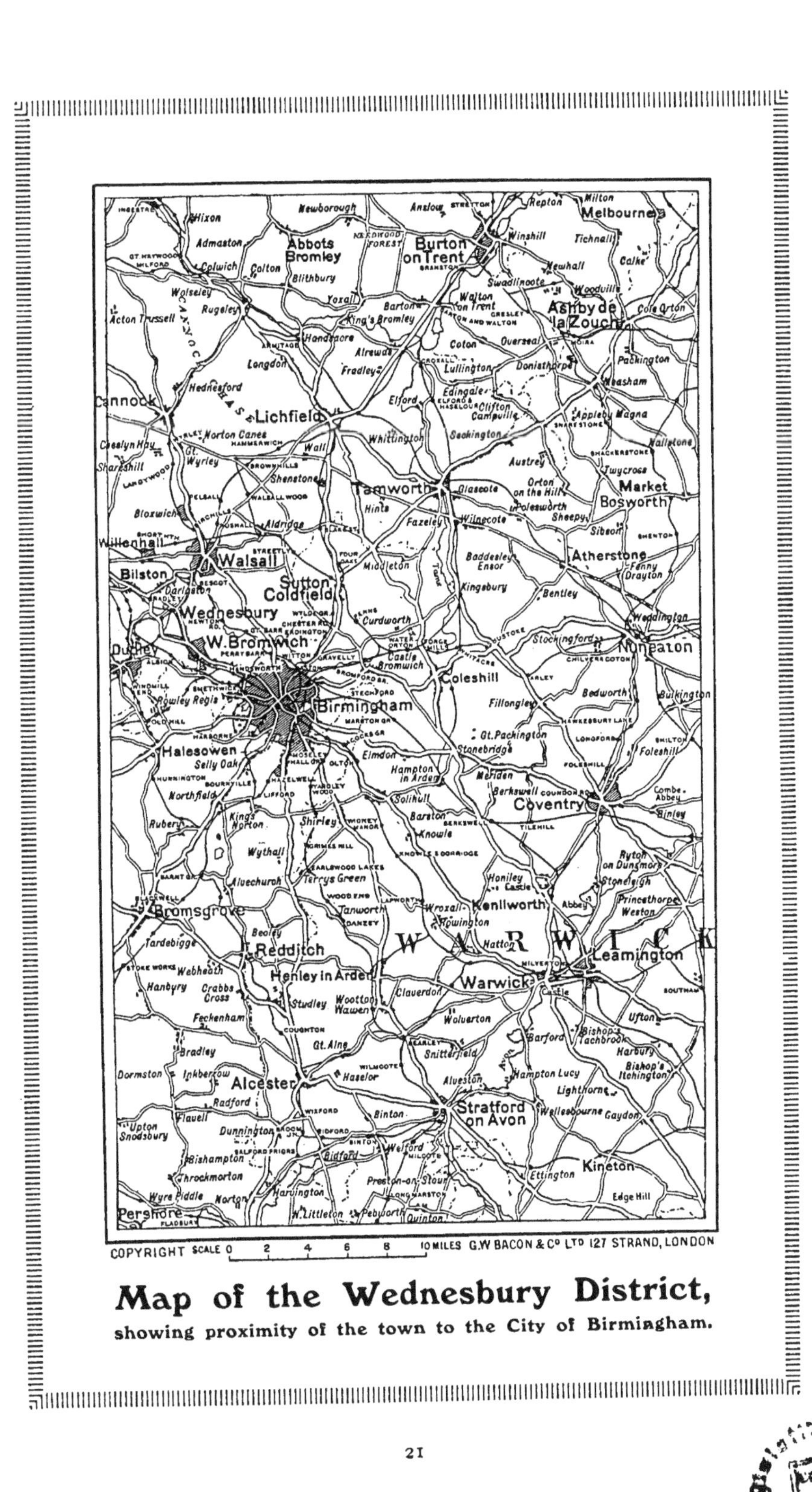

Map of the Wednesbury District,

showing proximity of the town to the City of Birmiagham.

Residential Wednesbury.

Population (estimated): 30,000.

Area of Borough: 2,287 acres.

No. of Houses: 6,108.

Death Rate (average last six years): 16·3 per 1,000.

Rateable Value of Borough: £106,060.

Post Office, Holyhead Road. Open week-days from 9 till 1 and 2.30 to 7; Sundays (for stamps, telegrams and registered letters only), 8.30 till 10 a.m. Telephone Exchange in Lower High Street.

Labour Exchange and Local Office for Unemployment Insurance, 48, Lower High Street. Office hours: 9 a.m. to 5 p.m., Monday, Tuesday, Wednesday and Thursday; 9 a.m. to 7 p.m., Friday; 9 a.m. to 1 p.m., Saturday. Hours for making claim for unemployment benefit and signing the register: Monday to Friday, 9 a.m. to 12 noon, and 2 to 4 p.m.; Saturday, 9 a.m. to 12 noon. Separate departments for women and girls.

Early Closing: Thursday.

Public Buildings.

The Town Hall and Municipal Buildings are situate in the Holyhead Road. The Hall seats about 1,200 people. In the year 1913 the buildings were much enlarged, the site of the former Education Offices being taken in.

The Art Gallery, adjacent, was erected in 1891, and contains a fine collection of pictures. It is connected with the Town Hall, and is used in conjunction with that building for big social functions.

The Public Library was erected in 1907-8, the sum of £5,000 being contributed by Mr. Andrew Carnegie, the site having been presented by a deceased Mayor and Mayoress of the borough (Ald. and Miss Handley). (See page 27.)

The Public Baths. The first Public Baths at Wednesbury were erected in 1878, but were re-constructed and brought up-to-date in 1913. They are situated in Walsall Street, about four minutes' walk from Upper High Street.

In Bridge Street is a commodious **Drill Hall**, now, of course, entirely in the hands of the Military, but formerly available for concerts and entertainments.

Places of Indoor Amusement.

The Hippodrome, formerly the Theatre Royal, in Upper High Street, seats between 1,400 and 1,500 people, has an excellent stage, and is visited by good Companies.

Also licensed for stage plays, music and dancing, **the Town Hall** is the scene of frequent entertainments in the winter.

J. W. Bernard, Photo.

The Science School.

There are two modern and handsome Picture Palaces—the Picture Palace, High Street, and the Picturedrome, Walsall Street—and an adapted house known as the Borough Theatre, in Earps Lane.

There are also the Wednesbury Borough Brass Band, the Crown Tube Works Brass Band, and the Banjo and Mandoline Orchestra, whose performances are much appreciated.

Schools.

In addition to six Church Schools there are four Council Schools within the borough, the largest and most modern being the Holyhead Road Schools, which accommodate 1,200 scholars.

Clubs and Institutions.

The Conservative and Unionist Club in Walsall Street has a large and well-appointed club house, which was re-built in 1904 at a cost of £3,000. The Liberal Club is comfortably housed in an adapted residence in Church Street. The Woden Club, Russell Street, non-political and unsectarian in character, was opened in 1906 for the social intercourse of those who could afford to pay a higher subscription than those charged at either of the two political clubs.

The Young Men's Christian Association in 1903 acquired High Street House for £1,500, where they now have very commodious and convenient premises, well furnished and equipped in every respect as a club house.

Flourishing Clubs are also connected with several centres of religious activity. The Parish Church Young Men's Social Club is situated in Brunswick Terrace. St. Paul's Institute and Parish Room at Wood Green has its home in large and specially-built premises, erected in 1906 at a cost of £2,500. There is a parochial Institute conducted in connection with St. James's Church; and a similar institution is associated with the Baptist Church in Holyhead Road.

The Wednesbury Building Society is an important institution, and has done much to assist the working classes in the borough to build or purchase their own homes, and, incidentally, to add to the housing accommodation of the town.

Public Market & Open Spaces for the Workers.

An open market is held every Friday (in addition to the usual Saturday night market) for the sale of all kinds of foodstuffs, general provisions and pedlary.

Brunswick Park, at Wood Green, is an example of what can be done by faith and well-directed industry. Twenty-eight acres in extent, the greater part was—thirty-two years ago—pit mound, whereas to-day its stretches of turf, well grown shrubs and trees, winding paths and beautiful flower-beds help to form one of the most charming spots in the whole of the Black Country.

I. W. Bernard, Photo.

The New Council Schools.

A similar Park—on a smaller scale—has been developing at Kings Hill during the past eighteen years, and bids fair to attain to the same standard of perfection as the Brunswick Park.

Open spaces in other parts of the borough have been secured for the recreation of the young people of the town.

Sports and Pastimes.

Wednesbury Football Charity Association, founded in 1880, arranges an annual Cup Competition, the proceeds of which are divided among the charitable institutions of the town.

The Wednesbury Old Athletic Club is primarily an old-established football team, with a playing ground in Lloyd Street, and its headquarters at the Crown and Cushion Inn, High Bullen.

The Amateur Football Club play on the Oval, and have their headquarters at the Horse and Jockey Hotel, Wood Green.

Wednesbury Cricket and Tennis Club have their headquarters at the Horse and Jockey Hotel, and their ground opposite thereto.

Wednesbury Town Bicycle Club, established 1884, have headquarters at the Talbot Hotel, Market Place.

Wood Green Cycle Club meet at the Cottage Inn, Wood Green.

The headquarters of the **Wednesbury Golf Club** are at the Talbot Hotel, Market Place, and the 9-hole Links are in Riddings Lane, about five minutes' walk from either station. Entrance fee: One Guinea; subscription, £1 11s. 6d. Terms for visitors: 1/- per day (Sundays and Bank Holidays, 2/-), 4/- per week, 10/- per month. Sunday play without caddies.

There is no fishing to be had in the immediate district, but there are excellent facilities for **Swimming** at the Baths in Walsall Street, where polo matches of an exciting nature are arranged from time to time.

Churches and Chapels.

The Parish Church of St. Bartholomew, which crowns the hill upon the slopes of which the town has been built, has already been mentioned. It contains an interesting picture representing the "Descent from the Cross," by Jean Jouvenet, a French artist of the seventeenth century. All the windows are adorned with stained glass, by Kemp, of London, and present a most handsome appearance.

St. John's Church, in Lower High Street, is a neat edifice standing in a well-kept churchyard; **St. James's Church,** off the Holyhead Road, stands amidst a forest of smoke-stacks, being situated in that quarter of the town studded with forges and ironworks. On the outskirts of the town are **All Saints' Church,** at Moxley, **St. Andrew's Church,** at King's Hill, and **St. Paul's Church** in the more salubrious

(*Continued on p. 30.*)

J. W. Bernard, Photo.

The Carnegie Free Library.

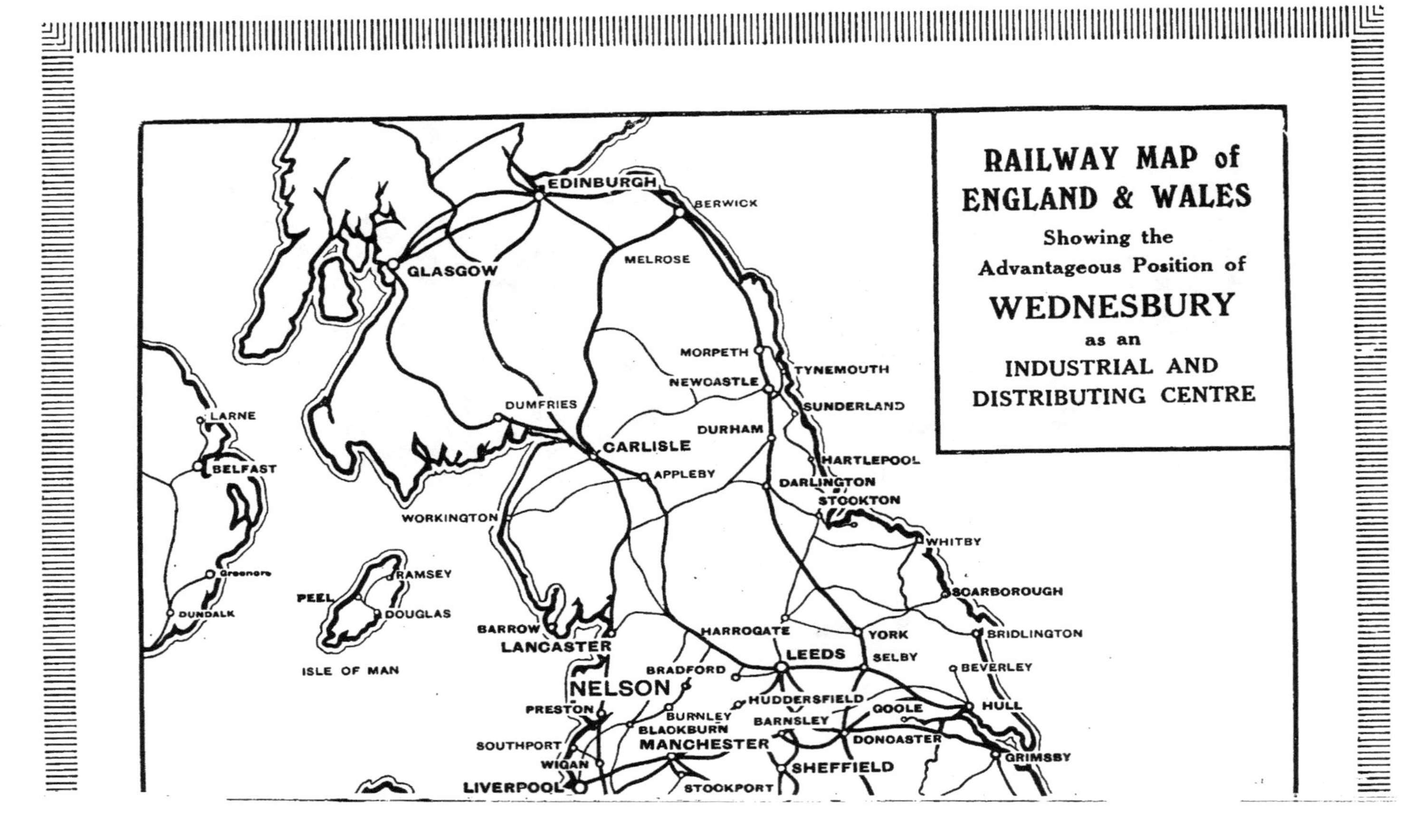
RAILWAY MAP of
ENGLAND & WALES
Showing the
Advantageous Position of
WEDNESBURY
as an
INDUSTRIAL AND
DISTRIBUTING CENTRE
EDINBURGH
BERWICK
GLASGOW
MELROSE
MORPETH
TYNEMOUTH
NEWCASTLE
SUNDERLAND
DUMFRIES
LARNE
DURHAM
CARLISLE
HARTLEPOOL
BELFAST
APPLEBY
DARLINGTON
STOCKTON
WORKINGTON
WHITBY
RAMSEY
Greenore
SCARBOROUGH
PEEL
DOUGLAS
DUNDALK
BARROW
HARROGATE
YORK
BRIDLINGTON
LANCASTER
LEEDS
SELBY
ISLE OF MAN
BRADFORD
BEVERLEY
NELSON
HUDDERSFIELD
PRESTON
BURNLEY
GOOLE
HULL
BLACKBURN
BARNSLEY
SOUTHPORT
MANCHESTER
DONCASTER
GRIMSBY
WIGAN
SHEFFIELD
LIVERPOOL
STOCKPORT

BARMOUTH
SHREWSBURY
WOLVERHAMPTON
BIRMINGHAM
LEICESTER
LYNN
YARMOUTH
PETERBOROUGH
ELY
RUGBY
HUNTINGDON
ABERYSTWITH
NORTHAMPTON
CAMBRIDGE
BURY ST EDMUNDS
WORCESTER
WARWICK
BEDFORD
BRECON
HEREFORD
BANBURY
IPSWICH
HARWICH
CHELTENHAM
GLOUCESTER
OXFORD
MERTHYR
PEMBROKE
SWANSEA
PONTYPOOL
NEWPORT
BRISTOL
SWINDON
LONDON
READING
CARDIFF
BATH
BARNSTAPLE
DOVER
SALISBURY
TAUNTON
BUDE
SOUTHAMPTON
BRIGHTON
NEWHAVEN
EXETER
PLYMOUTH
HG

atmosphere of Wood Green. The Mission Church of St. Luke serves, for the present, the populous suburb of Mesty Croft; and a mile or more well out of the din of the town is a small Chapel-of-Ease at the Delves, where lies a piece of open common land which some day will be more esteemed as an available breathing space than it is at the present time.

St. Mary's Roman Catholic Church is a handsome modern brick building, in the Early English style, the spire rising beside that of the Parish Church on the hill.

The two principal **Wesleyan** Churches are that in Spring Head, already mentioned, and the other in Holyhead Road. A third large chapel is situated at King's Hill, and there are several smaller ones in other parts of the parish.

Trinity **Congregational** Church, in Oakeswell End, is an edifice to which the denomination migrated in 1905, after being established upwards of a century in Russell Street. The **Baptist** Church, in the Holyhead Road, is the centre of much religious and social activity. The **Primitive Methodists** have chapels and considerable establishments attached thereto in Camp Street and at Lea Brook.

The **United Methodist Church**, in Ridding Lane, is a thriving community; it was the scene of a serious accident during the Anti-Romanist Murphy Riots of 1867. The Christadelphians and other minor sects have meeting-places in the town; the Salvation Army (who bought up the old Theatre Royal, Earps Lane, in 1882), have their "Barracks" in Upper High Street; and a Labour Church has been established at the old Quakers' Meeting House, in Lower High Street. Since 1897 a Free Church Council has federated the interests of the leading Nonconformist bodies of the town.

Useful Memorials.

The Clock Tower in the Market Square was erected to commemorate the Coronation of H.M. King George V., June, 1910. Close to Brunswick Park is a Nurses' Home, serving as a memorial of Queen Victoria's Diamond Jubilee—1897-8.

Banks.

London City and Midland Bank Limited, Market Place. Hours: 10 to 3; Saturdays, 9 to 12. Head Office: 5, Threadneedle Street, London, E.C.

London City and Midland Bank Limited (Metropolitan Bank Branch), Bridge Street. Hours: 10 to 3; Saturdays, 9 to 12. Head Office: 5, Threadneedle Street, London, E.C.

Lloyds Bank Limited, Holyhead Road. Hours: 10 to 3; Saturdays, 9 to 12. Head Office: 71, Lombard Street, London.

The Clock Tower, Market Place.

J. W. Bernard, Photo.

HISTORICAL NOTES.

Wednesbury is a place of considerable antiquity; as the original form of its name, Wodnesbeorge, indicates—it was a " beorh " or hill, dedicated by the Saxons to their war god, Woden. Two battles were waged here in those early times. The Anglo-Saxon Chronicle records that in the year 592 " there was a great slaughter in Britain, and Creawlin was driven out," this defeat of that King of Wessex by the Britons being located at " Wodnesbyrg." The second battle on this spot took place in 715, when Ceolred, King of Mercia, encountered his great rival, Ina, King of Wessex. In 914 Wednesbury was fortified against the Danes and the Welsh by the Princess Ethelfleda, a daughter of Alfred the Great, and at that time acting as Viceroy of Mercia for her brother, Edward the Elder.

The only traces remaining of that ancient fortification, which like all Saxon " castles " was of earthwork and massive timbers, is a portion of the western escarpment falling away within a few feet of the tower of the Parish Church; that sacred edifice having been subsequently erected on its site—the same site that in the pagan days of our progenitors had been occupied by a temple of Woden, an historic subject illustrated on one of the stained-glass windows of the church. In Norman times Wednesbury included Bloxwich and Sheffield, now outlying parts of the neighbouring town of Walsall; and until Henry II. exchanged it for an Oxfordshire estate, near to his favourite residence, Woodstock, was one of the Royal Manors. During those early centuries the boundary of the Royal Forest of Cannock passed right through the centre of the town, so that at least half its inhabitants were subject to the cruel forest laws of that barbarous period. In 1505 William Paget was born at Wednesbury. He was the son of a nailer, and rose to be a Secretary of State, an Ambassador, a Knight of the Garter, and one of the executors of the will of Henry VIII. He was astute as he was clever, and managed to flourish through all the treacherous changes of the Reformation period, as the friend and counsellor of the three succeeding monarchs, Edward VI., Mary and Elizabeth. He was the founder of the Marquis of Anglesey's family; his portrait, painted by Holbein, is extant.

During the later centuries no incident worthy of note occurred, although the mineral and industrial resources of the town were being slowly developed, as already recorded. In the old days of ill-regulated labour, the colliers and nailers, the gun-lock filers and gun-barrel makers, the blacksmiths and forgemen of " Wedgebury " (as the town was then vulgarly called) were sunk in ignorance and depravity, delighting in cock-fighting and bull-baiting, and in all the other brutal sports of the period. It was by a mob of these poor degraded creatures that John Wesley was assaulted in 1744, the Wednesbury Shrove-tide Riots of that year being one of the most historic incidents in the annals of Wesleyanism, and the subject of a fine painting by the late Marshall Claxton. The old horse-block from which the eighteenth-century evangelist used to preach in the High Bullen is preserved in front of Spring Head Chapel, the principal centre of Wesleyanism in Wednesbury—but John Wesley's first " meeting-house " in the town was in the locality now called Meeting Street on that account, and in a building which had formerly been a notorious cock-pit frequented by all the "sportsmen" for miles around. "Wedgebury Cocking" is the name of a coarsely-worded descriptive ballad of that period.

For an ancient town, Wednesbury is singularly devoid of old buildings and public foundations; probably attributable to the negligence and laxity of the long period of vestrydom. The Manor House stood to the north of the church, but only a few vestiges of it remain; blast furnaces now stand on the site of Willingsworth Hall; the most interesting building still standing is **Oakeswell Hall,** a half-timbered residence with a curious lantern roof, once the seat of the Hopkins family, who suffered for their loyalty to Charles I. during the Civil Wars.

The Parish Church, which has been extensively restored, is an edifice in the Perpendicular style, dedicated to St. Bartholomew. The earliest portion dates from the time of King John, when it was subject to Hales Owen Abbey. Among the memorials in the church are a mural tablet to the noble family of Harcourt, and a monumental tomb with the recumbent effigies of Richard Parkes and his wife, of Willingsworth, ancestors of Lord Dudley. There are also tablets to the Hopkins' family in Oakeswell Hall; to the family of Comberford (1559), who were Lords of the Manor; to Francis Wortley (1631); and there is the fragment of a slab, dated 1521, belonging to the Jennings family of litigious fame.

J. W. Bernard, Photo. **The Drill Hall, Wednesbury.**

UNDER GOVERNMENT CONTROL.

JAMES RUSSELL & SONS

LIMITED:

REGISTERED OFFICES: **Crown Tube Works, WEDNESBURY.**

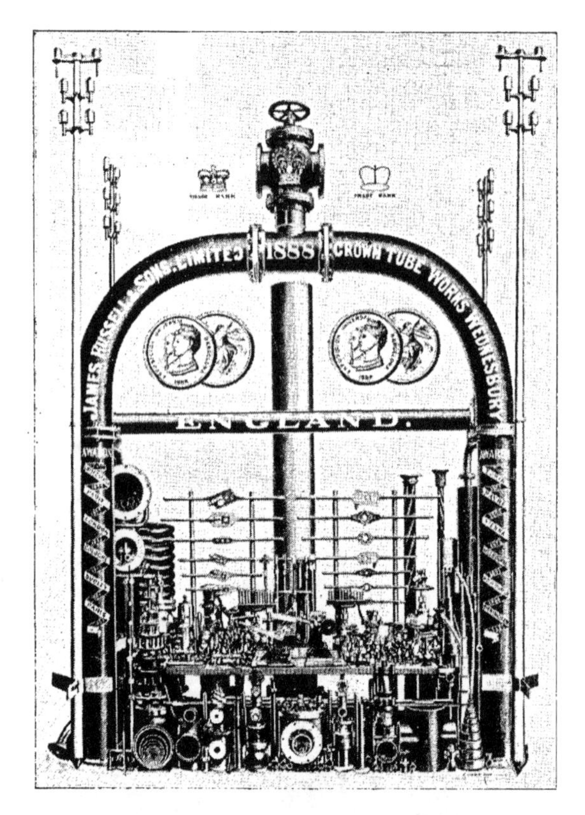

Inventors, Patentees, and First Manufacturers of

GAS TUBING.

London Office: 60, Queen Victoria St., E.C.

'PHONE: Nos. 2 & 67 Wednesbury. TELEGRAMS: "Russell, Wednesbury"

DEPOTS: London, Manchester, Leeds, Birmingham.

ıtsampto, Cannon, London.
B C Code, 5th Edition.

Cannon Street Buildings
(139, Cannon Street), E C.

THE STEEL NUT & JOSEPH HAMPTON, LIMITED
WEDNESBURY, ENGLAND

ALSO AT LONDON, GLASGOW, PARIS, BRUSSELS, MELBOURNE, SYDNEY, CAPETOWN, PORT ELIZABETH AND JOHANNESBURG.

Steel Dept.

Bright Drawn Steel Rounds, Squares, Hexagons, &c., for Machining; also in Shafting Quality.

Foundry.

Castings in Ordinary and Malleable Iron, &c. ∵ Tramway Castings a speciality.

Tool Department.

Lifting Jacks, Tube Cutters, Parallel Vices, Cramps and other Joiners' and Engineers' Tools.

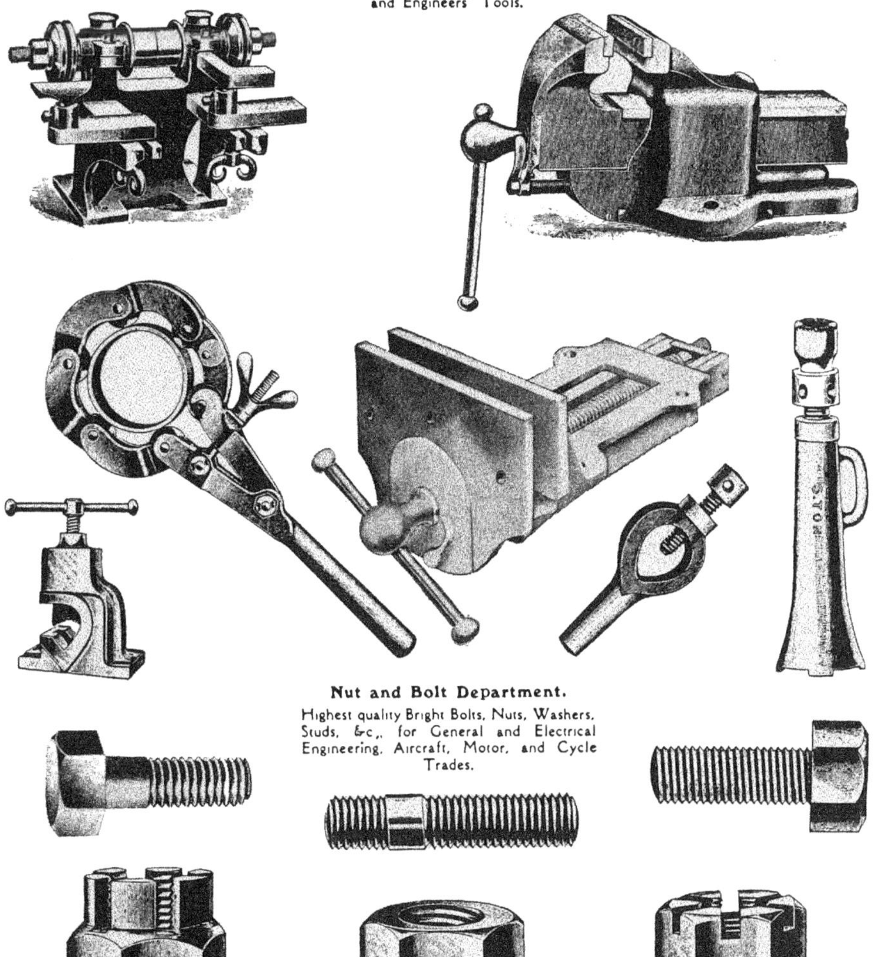

Nut and Bolt Department.

Highest quality Bright Bolts, Nuts, Washers, Studs, &c,, for General and Electrical Engineering, Aircraft, Motor, and Cycle Trades.

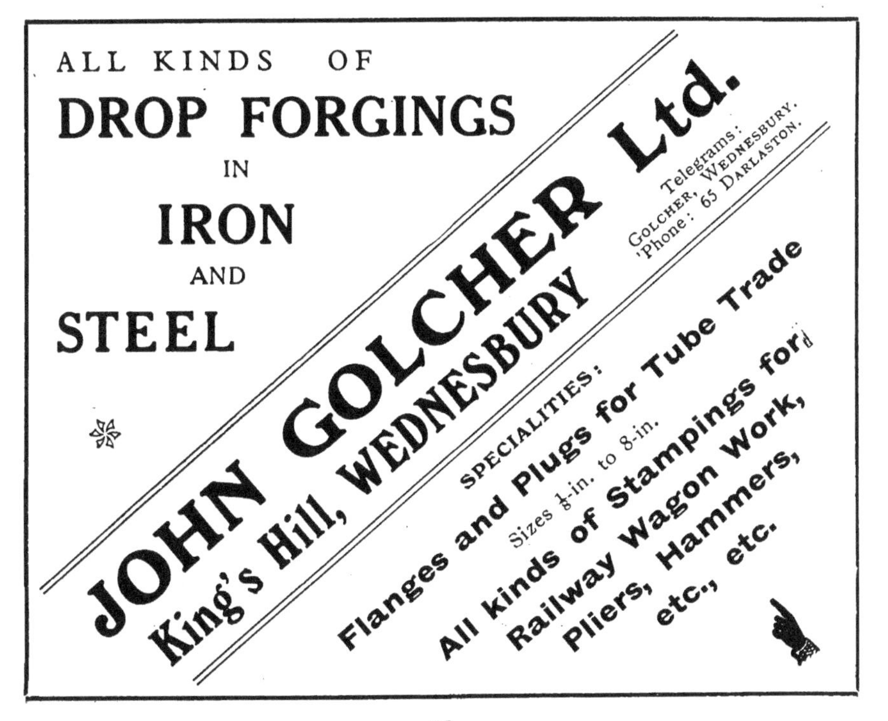

ALL KINDS OF
DROP FORGINGS
IN
IRON
AND
STEEL
JOHN GOLCHER Ltd.
King's Hill, WEDNESBURY
Telegrams: GOLCHER, WEDNESBURY.
'Phone: 65 DARLASTON.
SPECIALITIES:
Flanges and Plugs for Tube Trade
Sizes $\frac{1}{8}$-in. to 8-in.
All kinds of Stampings for
Railway Wagon Work,
Pliers, Hammers,
etc., etc.

BUSINESS DIRECTORY AND INDEX.

Style of Business.	Address.	Telegrams.	Telephone No.	Advt. Page.	Notice on Page
Axle, Wheel and Spring Manufacturers.					
Adams & Richards	Bridge Street, Wednesbury		8	41	16
Disturnal & Co., R.	Bridge Works, Wednesbury	"Disturnal, Wednesbury"	10	40	16
Richards & Sons Ltd., Edwin	Portway Works, Wednesbury	"Richards, Wednesbury"	98	39	15
Taylor (Springs) Ltd., A. H.	Bridge Street, Wednesbury	"Springs, Wednesbury"	48	38	15
Bolt, Nut, Stud and Screw Manufacturers.					
Barlow & Co., H. J.	Works: Mounts Road; Offices: Bridge Street, Wednesbury	"Barlow, Wednesbury"	161	45	19
Beebee, Alfred	Wood Street Works, Fallings Heath	"Beebee, Fallings Heath, Wednesbury"	81 Darlaston	2 cover	
Darlaston Bolt and Nut Co. Ltd. (Bagby & Harper)	Tower Bolt and Nut Works, Darlaston	"Bolts, Darlaston"	42 Darlaston	50	
Hampton & Beebee Ltd.	Wednesbury	"Washers, Wednesbury"	80	44	18
The Steel Nut and Joseph Hampton Ltd.	Wednesbury. (Agencies all over the world)	"Franchise, Wednesbury"; Code: A.B.C., 5th Edition	27 (two lines)	43	18
Castings in Iron and other Metals.					
Edmunds, W. P. (Light Grey Iron.)	King's Hill, Wednesbury	"Edmunds, King's Hill, Wednesbury"	78 Darlaston	51	
Peace Ltd., Henry (Phosphor Bronze and Gun Metal.)	Wednesbury		82 Wednesbury	50	18
Chemical Manufacturers.					
Chance & Hunt Ltd.	Head Office: Oldbury, near Birmingham; Works at Oldbury, Wednesbury and Stafford	"Chemicals, Oldbury"	203 Oldbury (10 lines)	49	20

BUSINESS DIRECTORY AND INDEX—continued.

Style of Business.	Address.	Telegrams.	Telephone No.	Advt. Page.	Notice on Page
Coach Ironwork.					
Taylor (Springs) Ltd., A. H.	Bridge Street, Wednesbury ..	"Springs, Wednesbury"	48	38	15
Electricity Supply.					
Wednesbury Corporation ..	Works and Office: Camp Street, Wednesbury	.. ..	.. ..	48	8
Engineers and Metal Workers.					
Edmunds, W. P... .. .. .. (Iron and Brass Founder.)	King's Hill, Wednesbury	"Edmunds, King's Hill, Wednesbury"	78 Darlaston	51	
Golcher Ltd., John .. .. .. (Drop Forgings.)	King's Hill, Wednesbury	"Golcher, Wednesbury"	65 Darlaston	51	
Peace Ltd., Henry .. .. .. (Brass Founders, Electrical and Mechanical Engineers, etc.)	Wednesbury	.. ..	82 Wednesbury ..	50	18
Platt Ltd., Samuel .. .. .. (Manufacturers of complete Plant for the Tube Trade, also Shafting, Pulleys and General Power Transmission Machinery.)	King's Hill Foundry, Wednesbury	"Platt, Wednesbury"	60 Wednesbury ..	42	17
Wesson & Co. Ltd., W. .. .. (Iron and Steel Bars and Strip, etc.)	Victoria Iron and Steel Works, Moxley, near Wednesbury ..	"Iron, Wednesbury"	62 Wednesbury ..	46	18
Galvanizers.					
Frost & Sons (Moxley) Ltd.. (Specialists in Tube Galvanizing by patent process.)	Falcon Works, Moxley, Wednesbury: also at Rough Hills Galvanizing Works, Wolverhampton	"Frost, Wednesbury"	72 Wednesbury 719 Wolverhampton	47	19
Wesson & Co. Ltd., W. .. .. (See "Engineers and Metal Workers")	Victoria Iron and Steel Works, Moxley, near Wednesbury..	"Iron, Wednesbury"	62 Wednesbury ..	46	18

Gas and Steam Cock Manufacturers.					
GARBETT, RICHARD	Portway Road, Wednesbury			41	
HAMPTON & BEEBEE LTD.	Wednesbury	"Washers, Wednesbury"	80 Wednesbury	44	18
Haulage Contractors.					
ADAMS & RICHARDS	Bridge Street, Wednesbury		8	41	16
Industrial Research.					
BIRMINGHAM CORPORATION LABORATORIES	Gas Department, Council House, Birmingham		Central 7016	3 cover	9
Iron and Steel Merchants.					
ADAMS & RICHARDS	Bridge Street, Wednesbury		8	41	16
Motor Accessories Manufacturers.					
DISTURNAL & CO., R. (Head and Hood Fittings, Door Hinges, Locks, Handles, Seats, etc.)	Bridge Works, Wednesbury	"Disturnal, Wednesbury"	10 Wednesbury	40	16
(See "Axles, Wheels & Springs.")					
TAYLOR (SPRINGS) LTD., A. H. (Motor Springs a Speciality.)	Bridge Street, Wednesbury	"Springs, Wednesbury"	48	38	15
Shafting Collar Manufacturers.					
HAMPTON & BEEBEE LTD.	Wednesbury	"Washers, Wednesbury"	80 Wednesbury	44	18
Stamping for Railway Wagon Works, etc.					
DISTURNAL & CO., R.	Bridge Works, Wednesbury	"Disturnal, Wednesbury"	10 Wednesbury	40	16
EDWARDS LTD., JOB	Eagle and Junction Works, Wednesbury		36	35	
GOLCHER LTD., JOHN	King's Hill, Wednesbury	"Golcher, Wednesbury"	65 Darlaston	51	
Timber and Slate Merchants and Roofing Erectors.					
WALSH GRAHAM LTD., C	Wednesbury, Wolverhampton and Smethwick	"Graham, Potters Lane, Wednesbury 12"	12 Wednesbury	52	19
Tool Manufacturers (for Joiners and Engineers).					
THE STEEL NUT AND JOSEPH HAMPTON LTD.	Wednesbury. (Agencies all over the world.)	"Franchise, Wednesbury"; Code: A.B.C., 5th Edition	27 (two lines)	43	18

DIRECTORY AND INDEX—continued.

ADDRESS.	TELEGRAMS.	TELEPHONE NO.	ADVT. PAGE.	NOTICE ON PAGE
	hise Wednesbury"; B.C., 5th Edition..	27 (two lines) ..	43	18
	.. ..	36 Wednesbury ..	35	
	.. .. ..	.. ..	6	13
	y"	.. ..	37	
	..nesbury" ..	31	6	13
	hard, Wednes-	13 Wednesbury	4 cover	12
	bury"	2 & 67 Wednesbury	34	12
	.. ..	.. ..	36	
be	runswick, Wednesbury" ..	32 Wednesbury ..	37	14
	ubes, Wednesbury"	4 & 53 Wednesbury	10	13
	ashers, Wednesbury"	80 Wednesbury ..	44	18

Indus al Research

MANUFACTL ERS IN WEDNESBURY 'ND DISTRICT

Are cordially invited ail themselves of the facilities afforded by

THE BIRMINGHAM CORPORATION NDUSTRIAL RESEARCH LABORA- ORIES AT THE GAS DEPARTMENT, OU H USE, BIRMINGHAM.

se lave ben equipped with a view to collabo- ng cturersin carrying out work of a scientific inveigational character—

FOUNDY AND HEAT TREATMENT SHOPS

ed to enale manufacturers to carry out tests on a full manufacturing scale

WORK has already been carried out in connection with :—

; and Alling,—Annealing of Metals,—Heat Steel,—Md Finishes,—Glass Manufacture,— aste Materil—Industrial Heating Plant,—Auto- ng and Meta Cutting,—Testing of Ferrous and Metals and oer Materials,—Electrical Testing,— l TemperatuMeasuring and Recording Apparatus.

rers need hano hesitation in bringing to the work of a cfidential nature, and can rely in pon the utmoare being maintained

culars :—

NGINEE-IN-CHARGE,
trial Research oratories,
t., Counc House, Birmingham.

TELEPHONE - CENTRAL 7016.

BUSINESS DIRECTORY AND INDEX—continued.

STYLE OF BUSINESS.	ADDRESS.	TELEGRAMS.	TELEPHONE No.	ADVT. PAGE.	NOTICE ON PAGE
Tramway Castings' Specialists.					
THE STEEL NUT AND JOSEPH HAMPTON LTD.	Wednesbury	" Franchise, Wednesbury "; Code: A.B.C., 5th Edition..	27 (two lines) ..	43	18
Tube and Fitting Manufacturers.					
EDWARDS LTD., JOB	Eagle and Junction Works, Wednesbury	.. ..	36 Wednesbury ..	35	
GRIFFITHS & SONS, ISAAC	Imperial Tube Works, Mesty Croft, Wednesbury	.. ..	.. ..	6	13
GRIFFITHS & BILLINGSLEY	Victoria Tube Works, Mesty Croft, Wednesbury	" Victoria, Wednesbury "	.. ..	37	
McDOUGALL LTD., JAMES	Hope Patent Tube Works, Wednesbury	" McDougall, Wednesbury " ..	31	6	13
PRITCHARD LTD., THOMAS	South Staffs Tube Works	" Thomas Pritchard, Wednesbury "	13 Wednesbury ..	4 cover	12
RUSSELL & SONS LTD., JAMES..	Crown Tube Works, Wednesbury. London Office: 60, Queen Victoria Street, E.C. 4.. ..	" Russell, Wednesbury "	2 & 67 Wednesbury	34	12
RUSSELL & CO. LTD., JOHN ..	Old Patent Tube Works, Wednesbury. (Head Office: Alma Tube Works, Walsall)	.. ..	.. ..	36	
SMITH LTD., EDWARD	Brunswick Tube Works, Wednesbury..	" Brunswick, Wednesbury " ..	32 Wednesbury ..	37	14
SPENCER LTD., JOHN	Globe Tube and Engineering Works, Wednesbury. London Office: 16, Dowgate Hill, E.C.	" Tubes, Wednesbury "	4 & 53 Wednesbury	10	13
Washer Manufacturers (for the Aircraft, Motor and Engineering Trades).					
HAMPTON & BEEBEE LTD... ..	Wednesbury	" Washers, Wednesbury "	80 Wednesbury ..	44	18

TELEPHONE:
15 Wednesbury.

TELEGRAMS:
Thomas Pritchard, Wednesbury

Thomas Pritchar

Ltd.

MANUFACTURERS OF

WROUGHT-IRON TUBES AND FITTINGS
FOR

Gas, Steam and Wate

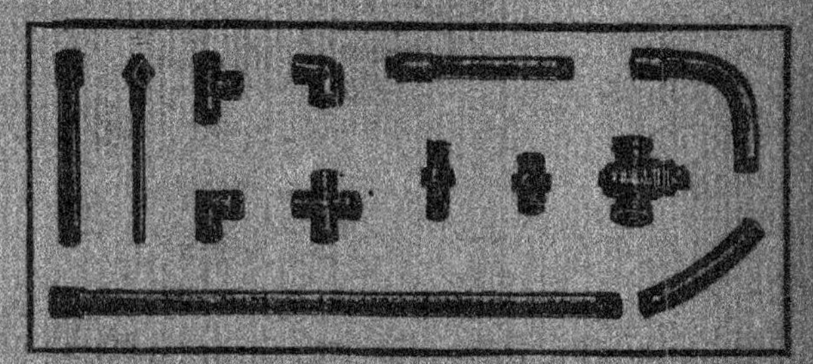

LAP WELDED TUBES FOR LOCOMOTI
AND MARINE BOILERS.

WELDED TUBES FOR BEDSTEADS AND BLINDS
CORE BARS, PLAIN AND TAPERED COILS, AN

TUBULAR LADDERS,
ALL GOODS OF FIRST-CLASS QUALI

South Staffs Tube Work

WEDNESBURY.

LONDON OFFICE:
29, Great St. Helen's St., E.C.

AGENCIES at
Liverpool, Birmingham, Bris

CPSIA information can be obtained
at www.ICGtesting.com
Printed in the USA
BVHW04*1205060818
523683BV00013B/236/P